Robotics

Discover The Robotic Innovations Of The Future

An Introductory Guide To Robotics

Dr. Kevin Klein

¹Table of Contents

1

Introduction

To put it in simple terms, robotics is the study or science involved with designing, building, theorizing about, or using robots. While there are many different fields that are involved in the end result of actually working with robots, those will not be discussed here. The more direct subjects that relate to robotics are the focus of this book.

A robot is a simple enough idea. It is a machine that can do something by itself, in the simplest terms. You have almost certainly seen them in movies, or read about them. Many people think of a metallic, human looking, machine when they think of a robot. The reality these days is a little less dramatic than that. In appearance, modern robots are often complex limbs or moving tools. They can complete tasks largely without the need for human assistance, but they are a long way from the types of robots people have been imagining for centuries; that's right—the idea of the robot is very old indeed.

The things that robots might be able to do, are a key driving force beyond their development. These ideas drive related fields, and those in turn drive people to come up with better ways to make and use robots. This cycle has been going on for some time, and it goes back a lot more than just a few decades (but you can learn more about that in the History of Robotics chapter).

Many people believe that robotics is currently the highest point of technological development. Once society has highly functional and mobile robots at their command, what could happen in the future?

What Is Robotics?

Robotics is a merger of varying scientific areas, which mainly uses the advances of the following fields:

- Manufacturing technology
- Material science
- Mechanical engineering
- Advanced algorithms
- Fabrication of sensors

Anyone who even dabbles with robotics, is going to find themselves looking into many different types of scientific study and practice—and by many, think *hundreds* of them. People often choose to begin with robotics due to some type of romantic notion about how much robots can do for the world, and humanity. Imagine how the world could change after the successful implementation of these wondrous machines; the likes of which have only been realized in science fiction books and movies. If you're thinking of getting started with robotics, get ready for a lifetime of adventure and seemingly magic possibilities.

As we have already discussed briefly, robotics is the study, design, theory, or use of robots. But that's such a general and shallow explanation, that delving deeper will be our next step. What makes a robot, a *robot?* That topic can easily turn into a philosophical debate, or a technical one, or both. For the purposes of getting an introduction to robotics, it's best to learn the most common definitions for now. Once you have a better understanding of the basics, you will be free to move on to more complex topics for study.

Mechatronic Devices

The terms "robot" and "robotics" are not one-and-the-same. A robot is a constructed thing that carries out any type of action or behavior automatically. Robotics is the theoretical and practical process behind designing, building, and applying robots. This can involve your garage door, which might open as you approach. The door would have its own sensor, which detects any signal that is sent to it from a remote. Once it detects the right signal, it would use its actuators to physically open the door, and it would have a controller to switch of its motors once the door it shut. This is technically a robot, although the term "mechatronic device" is more accurate. However, for our purposes to introduce the concept of a robot, it works just fine.

So, to reiterate, the main parts of a mechatronic device are:

- **Sensors**, to detect what is going on around the device.
- **Actuators**, to actually move and do something.
- **Control system**, to work the actuators while being aware of how they are affecting the surroundings, using the sensors as guides.

"True" Robots

So, if the mechatronic device we discussed above is not a proper robot, what is? It does have the same components, but there are also some extra types of parts required (this will be discussed in detail in the Key Components chapter of this book). In order to be a robot, the device must have autonomy, which can also be called resourcefulness. It has to be able to actually do things by itself, without needing a person to guide it along or control it. Can you see how the "automated" garage door in our previous example lacks that basic ability? You would need to press the button on your remote control in order to tell the door to open. That's why it is not really a robot.

Of course, some people will debate what makes a robot a robot. Many people would say that the garage door is a robot, because it can open by itself, even if that's only after a human sends it the signal to do so. But that's not really important right now.

These are two highly common types of robots:

- **Automated machines**, which are machines that have their own actuators and sensors to inform them how to use move. They can carry out some type of task without human guidance. This might be as simple as an automated vacuum robot; the type that is very popular in households these days. It could also be a cleaner, or a CNC machine for cutting construction material. The majority of commercial and industrial robots will be classified as automated machines (or autonomous machines, which has the same meaning).

- **Pet machines**, which are able to move around at least a little bit. They have sensors to learn what it going on around them, even though this might be extremely basic. They are also able to act upon what they sense. These types of robots do not really have much use, apart from entertaining people, or helping their creators understand robotics better. Making a robot pet is a great way to learn more about how robotics works, in a fun way.

You have probably seen machines that look like robots, but are not actually robots in any way. This includes the fighting robots that are very popular. While they are a great way to learn about how to design and built a lot of components that do go into robotics, they are not robots. They're remote controlled machines, much like radio controlled cars or planes that you might have had as a child. They can be pretty serious pieces of hardware, but they lack the autonomy to be robots, or the sensors to even be mechatronic devices.

Generally speaking, if a machine has its own autonomy in performing tasks, it's a robot. Otherwise, it might be very close, but it's not a robot.

History of Robotics

Let's go back to the what a robot is, just briefly. It's a machine that can be programmed to perform at least one task. It doesn't necessarily have to be like a person in its actions or appearance. However, since the idea was first conceived by people, they have been imagining just that.

The Term "Robot"

The actual word "robot" was first introduced in 1921. A Czech playwright named Karel Capek (1890 – 1938) is noted as being the first person to use this word. It might have been his brother who suggested this word to him, however. He was reported as being a Nobel prize candidate a number of times. This was because of his great volume of work as both a playwright and author.

Capek used the term in his very successful 1921 play, which was called R.U.R. The play was about a company called Rossum's Universal Robots, a company that made robots. In the play, people are living in a utopian world, thanks to the many wonderful things that the robots have brought about. While the robots bring a lot of happiness and a better way of life at first—unemployment levels eventually become a huge problem. You can see this sort of thing happening throughout history, as new technology is created that can do the work of people.

Early History

The idea of an automated helper goes back to the ancient mythology and folklore. The story of Pygmalion involved a statue coming to life. Cadmus was said to have planted the teeth from a dragon, which then formed into soldiers. And then there are the golems made from clay, which can be found in Jewish mythology. The Norse legends also involved giants made of clay. A Chinese account from around 1,000 BC included automated things that looked like humans.

The idea of automatons was very popular throughout medieval time, as can be noted in their literature. By the 1700s, people were actually created automated machines, and this field because common. People were amazed with the intricate designs that were being created. It was not until more modern times that anything like newer robotic machines, which we will be focusing on in this book, were made.

The modern robots that we see today, with their electronic parts and advanced programming, have some pretty ancient relatives. The early machines ancient people made are not completely different. Yes, a lot of these contraptions were mostly just novelties, but the amount of thought and scientific application that went into them might surprise you.

There were working humanoid automatons, as well as animals. These types of machines used systems of water, weights, or other low-tech methods, for power. There was also a huge industry of clockwork machine making after medieval times, with levels and springs being put in place. These allowed the devices to perform simple tasks. While these are pretty primitive by modern standards, they were ingenious at the time. They helped to give scientists a place to start in the field of robotics as well. And that knowledge has helped robots to advance to where they are today.

Isaac Asimov's Three Laws of Robotics

It might not be that much of a surprise to learn that science fiction authors have had a lot of influence over the core ideas that are used in real technology. After all, many of the world's leading inventors will attest to having a love for Star Trek, or something similar. Isaac Asimov (1919 – 1920) was a successful writer during the Golden Age of science fiction. You might have heard of his story, *I Robot*, and the other stories of the same setting. In it, he came up with three laws that dictated how a robot should behave. He added another law later, calling it the zero law.

- Law One: A robot may not injure a human being or, through inaction, allow a human being to come to harm.
- Law Two. A robot must obey orders given it by human beings except where such orders would conflict with the First Law.
- Law Three. A robot must protect its own existence as long as such protection does not conflict with the First or Second Law.
- Law Zero. A robot may not injure humanity, or, through inaction, allow humanity to come to harm.

These laws have been used, altered, or referenced, many times throughout the decades.

Unimate, the First Robot

As you might already be aware, there was a lot of technological advancement because of World War II. While war itself is clearly never a good thing, the drive to compete and get the upper hand over opponents, has brought about a lot of advanced machinery and ideas. It was in 1945 when George Devol (a successful entrepreneur and inventor) and Joseph Engelberger (an engineer) started to talk about Isaac Asimov's writing.

These two men decided to work together to create an actual robot. And they wanted to create one that would be both functional and commercially viable. They used their idea to convince Norman Schafler from Condec Corporation to back them.

Once they had the support they needed, Engelberger got to work by making a company called Unimation, which was a shortened version of the "universal automation". This was the very first commercial company that would make robots. While his partner was setting up the manufacturing side of things, Devol put in the patents that they would need. They created the first robot ever, which was given the nickname of "Unimate", after their company's name. This is why Engelberger is commonly known as the "father of robotics".

The first one of these Unimate robots went to work at a General Motors factory. It was put to work with die casting, as this was a job that people despised doing. Most of the Unimate robots also went on to do similar types of work, as well as spot welding cars in the factories. This was a huge success in a commercial sense. Robots were faster, cheaper to use than people, and quickly adopted by other companies. The field of robotics was no longer a purely scientific one; there was some clearly some serious money to be made by roboticists and their customers.

The Uses of Robotics (Applications)

If you have an idea of what a robot is, you probably have a lot of ideas about what they might be able to do. Robots have typically been used for any tasks that are either to **dirty, dull**, or **dangerous** for a human being to do themselves. If a human is willing and able to do a job, and it's cheaper to hire them, there's a good chance there won't be a robot designed to do it any time soon.

There are two types of robots used in real world application at this point in time:

- **Dedicated robots**, which are made to do a specific task, or a very small range of related tasks.
- **General purpose autonomous robots**, which are designed to do a range of different tasks by themselves.

General Purpose Autonomous Robots

These are made to carry out a variety of tasks. Quite often, they can find their way through spaces that are known to them, interact with things like doors and lifts, recharge themselves when they are running low on power, and perform other tasks. But these are often fairly simple tasks, that would be thought of as nothing by a human. General purpose robots can also link up to networks, like a computer could, so they can access more resources and become more useful; resources like extra software, better computing power, or accessories.

General purpose autonomous robots can often be found doing the follow tasks:

- Interacting with people through speech and motion
- Recognizing individuals
- Acting as company to humans
- Monitoring the quality of certain environments
- Responding when an alarm is triggered
- Collecting supplies
- Other menial tasks

The list above might not seem like anything special for a human. The thing about these robots is that they can carry out many different tasks, and swap between them without the need to be refitted, reprogrammed, etc. They can even spend a day going from one task to the next, much like a person would.

There are robots that are made to look and act like people; these are known as humanoid robots. This type of robot is still quite primitive, although they have been in the imagination of the human race for a very long time. Humanoid robots are still unable to carry out all of the aforementioned tasks without error, even when they're in familiar places.

Dedicated Robots

These are the true robot workers that are in use today. While the general purpose, or humanoid, type discussed above might be big in the future—it's the dedicated worker robots that are actually getting things done right now. You might not realize just how much of the world's work is already carried out by robots. It might even shock you to find out how many jobs are already being taken over by robots. If that type of thing, or anything else to do with robotics and the human race, does concern you, please don't forget to read the section about ethics in the Future of Robotics chapter.

Commercial Robot Uses

These are just some different jobs done by robot workers right now:

- **Electronics manufacture.** The massively produced type of printed circuit boards (PCBs) are made almost entirely by robots. They are a pick and place type of machine, which can take the electronic parts from little shelves and then put them onto the circuit boards. Robots can do this with extremely high accuracy. There is simply no way a human worker could hope to compete with their speed and output.

- **Material handling.** Manufactured products and their material are being handled, packaged, put on pallets, or handled in some other way, by robots all over the world. They are great at moving packaging material to where it needs to be, or working with loading and unloading machines. Around 38% of the working robots on the planet are being used to for some type of material handling purposes. This number is constantly growing, however.

- **Dispensing.** This type of work includes painting, spraying, gluing, sealing with adhesive, etc. Even though the robots can do a very accurate and smooth job, only around 4% of the world's working robots are used for dispensing related tasks.

- **Picking.** One company taking advantage of robotic workers is the massive Amazon. They are using small, box looking, robots to actually carry shelves of items to workers, so they can be scanned and packaged.

- **Assembly.** The jobs that robots carry out in the assembly field are, press fitting, fixing, disassembling, inserting, etc. However, as other robotic technology has been introduced, the use of robots has diversified. This has led to a decrease in the use of robots for assembly purposes. These new technologies include tactile sensors and force torque sensors, which give better sensory capabilities to the robots.

- **Car manufacture.** Robots have been used heavily in automobile production for the past few decades. A car factory usually has hundreds of individual robots in it now. These make up production lines that are completely automated. For every worker who has a job in a car factory, there are around ten robots. The robots are set up in different stations throughout the assembly line, and they can complete entire sections by themselves (that is, with no human help required).

- **Welding.** Arc or spot welding is the most common type done by robots. And just like the first commercial robots ever made, these robots are working mostly in the automotive industry. Even the smaller manufacturing workshops are starting to use robots for their production. It's getting a lot more feasible to do so, with the costs for robots coming down steadily.

- **Processing.** Processing robots carry out tasks like cutting with water or laser jets. This is just a small part of the robot industry, with around 2% being used for processing. It is most likely due to the fact that simpler automated machinery can be used for the same tasks.

- **Military Use.** The robots used by the military are some of the most advanced in the world. They are also considered the most important by a lot of people. They are the height of robotic technology, and plenty of people rely on that fact to remain safe and alive. It's important to note that the military's use of robots is a heated topic for political and ethical debate. Some of the robots commonly used by the military are:

- EOD. Explosive Ordinance Disposal robots are able to locate and deactivate mines and explosive devices, or bring them to someone who can examine them further.
- UAVs. Unmanned Aerial Vehicles are capable of flying without human pilots. They are used to gather information in dangerous areas, spy on enemies, look for hidden explosives, or provide a bigger view of battle areas.

The Benefits of Robots

Robots can give human workers, industries, and the human race, a lot of benefits. If they were to be properly introduced, commercial robots could greatly improve the overall quality of life on this planet. They could help to free workers from dull, filthy, hard, and dangerous jobs. While many people argue that this could also lead to mass unemployment and shortages for the majority of the labor force, that is all open to debate at this point in time.

Robots can help industry by giving better control over management, increasing the productivity of the workforce, and helping to ensure products and services that are consistently of a high quality. A robot is able to continue to work without growing tired, throughout the day and night. Apart from malfunctions and maintenance, they will not lower their performance due to wondering thoughts or fatigue.

All of this means that the use of robots in industry can greatly reduce how much things cost to buy. This would be of immense benefit to all countries, especially those that are unable to produce enough basics, like food and safe water, to provide for their populations. How the robotic workforces are actually used to benefit the human race is another topic, however? More on that will be covered in the Future of Robotics chapter.

Careers in Robotics

Are you interested in a career in robotics? You have probably realized that not everyone in this industry actually works directly with robots. There are many different fields that are all intertwined, and which feed off each other to create new technologies. And as those new technologies are made, they change things like power production, manufacturing, and research. This in turn creates a growing number of jobs for people who want to work in robotics, and related areas.

Modern robotics is largely focused on the moving and lifting of physical objects, as in the jobs discussed in the chapter titled, The Uses of Robotics. Robots are primarily designed to carry out fairly specific tasks at this moment in history. A grown human being would be able to walk over to a box, pick it up, and carry it to the required destination, all without thinking about their individual movements and decisions. For every action that a robot needs to do, there is a lot of analysis to be carried out.

Once all of the smaller actions that a robot must take are analyzed, they can be put into action with machinery and programming. Those tiny actions are eventually combined into a larger piece of machinery: the actual robot. It takes a lot of specialized workers to come up with all of those little things that go into a seemingly simple task, like having a robot move a box from one place to another.

There are the common jobs in the field of robotics:

Operator

If you want to actually work directly with functioning robots, without being in the research field, a job as an operator might be ideal. These types of jobs make up the bulk of the robotics industry's workforce. They include things like pilot technicians (who are electromechanical technicians), operating robotic vehicles like submarines, for example. They operate things like the systems for launch and recovery of robots. They also operate the tools that are attached to robots, where full automation is not in use. Operators will also help robotic engineers to design new equipment for robots, and improve them. Operators will also know how to repair the tools that they operate.

Robotics Technician

These workers are often supervised by robotics engineers. They work to make the systems that the engineers have designed, in a hands-on approach. They will also work through any problems with the designs, troubleshooting components and systems, or keeping documentation maintained. Technicians might also help with the construction or design of robotics. They might be required to incorporate what an engineer has come up with, or design a physical platform for a robotic machine to work with. They help to install robotic systems, hooking up wires and putting actuators into place, so that a robot can function.

Robotics Engineer

An engineer should have training in mechanical engineering, computer science, electronic engineering, electrical engineering, or a mixture of these fields. Whatever they specialize in, a robotics engineer needs to know how to work out problems with computer programs. They will need to understand the installation of robotics systems. They will often be in charge of other workers, who are building robots, ensuring that everything is in line with specifications. They overview and preview the designs of robotics technicians, testing and repairing any issues with systems.

Related Jobs

These are some of the indirect jobs related to robotics:

- Safety system installation
- Welding
- Maintaining and repairing robots
- Operating hydraulic equipment for tests
- Failure analysis tests
- Programming or reprogramming robots
- Interpreting schematics
- Installing or removing robotic equipment

In December 2014, it was reported that a robotics technician could earn $55,000 USD on average. Robotics engineers were earning around $75,000 USD as a median salary.

Key Components

There are a lot of different robot types, and they're used in many various environments. Even though they are used for so many different tasks and industries, they do have some basic things in common. This was just briefly covered in the What Is Robotics? chapter at the start of the book. We will now look at what makes up a robot, in much greater detail.

While these basic parts are all present in modern robots, you must remember that there are thousands of different parts that can go into just one component.

Three Key Aspects of a Robot

Form

A robot will have some type of physical form, which is a mechanical construct. This is generally a type of frame, or some other shape that helps it to carry out a certain task. A robot that is designed to travel along a road might have some type of appropriate wheels. The type of form used will change, depending on the intended function of the robot. Robotic engineers will create what they need to, in order to get a specific job done. You will find some truly odd looking robots because of this. However, there are plenty of very boring looking robots, which have been designed to be practical and useful.

Electrical Components

In order to be able to do anything, a robot needs to contain electrical components. These are used to control the machinery of the robot, and to provide power to it as well. The robot in the previous example, which needs to move down a road on its wheels, would require power to turn those wheels. This is almost certainly going to be in the form of electricity. This will need to go through wires inside of the robot, and come from a battery, which stores the electricity.

Even if a machine is powered by petrol, it will need to use electricity to start a combustion engine; that's why cars and other engine powered machines need to have batteries. Other electrical aspects of a robot include motors for movement, and sensors for monitoring what is going on around the robot.

Computer Code

All robots need some computer programming in order to be able to "think". Without this code, they would not be able to decide what they should do next, or if they should act at all. In short, they would not be robots in the true sense of the word, but simply mechatronic devices. On the other hand, many non-robot machines still have computer code.

Looking at our example of the wheeled robot, we know that a properly constructed form is required, as well as power to provide movement and sensory capability. However, it would not be able to move automatically, without computer programming code to tell it to do so.

Programming is the "brains" behind robots. Without a good program, even the best robot design will not function as it should.

Here are the three types of robotic programs:

- **Remote control.** This is not the same as a traditional remote control, which is used to control a machine in a step-by-step way, by a human. Remote control programming consists of commands that are programed in advance. When the robot gets the signal, it will carry out what it was programmed to do. The signal is usually given by a human, who can then do something else. Robots that use this type of programming may be thought of as automations, but not for the purposes of this book.
- **Artificial intelligence.** In the purest sense of the term, many people think of AI robots when they think of them at all. This is largely because of what movies and books have showed us throughout the decades. A robot that has an artificial intelligence program will be able to interact with things around them, without requiring a signal from any control source. They can decide how to react to occurrences and objects, using their programming code to tell them what they should do.
- **Hybrid.** This is a combination of remote control and artificial intelligence programming.

Further Components

Sensors

These are what lets a robot gain information about what's around them: their environment. They also allow a robot to be aware of what's going on with their own internal parts. To be able to do any of the things required of a robot, sensors are essential. They allow a robot to detect changes around them, so they can take appropriate action. The robot that needs to drive down a road on wheels, from our earlier example, must be able to detect changes in the surface of the road, or obstacles that might be in the way.

Sensors are also used as a part of safety features, or to warn the robot and its operators about malfunctions. This real time information is what lets the robot have an idea of where it is, and what it is doing.

Touch and vision are two of the primary sensor types being developed at the moment. Other common types include radar, lidar, and sonar.

Actuators

You can think of these as a robot's "muscles". They are the component that takes stored up energy, and translates that into motion. Electric motors are the most commonly used type of actuator, by a lot. They work by turning a gear or wheel around to provide movement (as with your car). Linear actuators are also used a lot, and mainly for robots in factories.

Different types of actuators include:

- **Electric motors.** DC motors are often used in mobile robots, while AC motors are used for industrial ones that can remain in one place and do their job. These are best used where some type of rotation is required, and when extremely heavy loads are not being dealt with.

- **Linear actuators.** There are various types of these. They are able to quickly change directions, going in and out. They are great for use with heavy loads, and are typically moved by compressed air, or oil.

- **Shape memory alloy.** This is also called "muscle wire". When it is powered by electricity, it is able to contract, and then return to normal shape once the electricity is removed. Muscle wire is ideal for small robotic parts, because of its simplistic, lightweight design.

- **EAPs.** These are electroactive polymers, which is a type of plastic that's able to contract as much as 380%. As with "muscle wire", it does this with electricity. EAPs have been used in the arms and even faces of robots designed to look like humans. They can also be used to give robots the ability to walk, fly, float, or swim.

- **Piezo motors.** These are a newer option, where DC motors have traditionally been used. While they can be used for the same applications, they work quite differently. Small piezo ceramic elements are used to create thousands of vibrations each second. This in turn causes either rotary or linear motion. Some of the ways that piezo motors are used include vibrating a nut, screwing in a screw, or moving a motor in a typical way. Why would someone want to use something this complex, when electric motors are already available? Piezo motors can provide a lot of force and speed for how tiny they can be made.

- **Pneumatic artificial muscles.** These are also known as "air muscles". They are expandable tubes, which are moved by air.

- **Series elastic actuators**. These involve a spring incorporated with a motor actuator.

- **Elastic nanotubes**. These are only being newly developed, but there is a lot of promise in these actuators. Because there are no defects in carbon nanotubes, they can store a lot of energy when moved. Just an 8mm strand of the material could be used to do the same job as a human being's bicep. This type of "muscle" power could allow robots to be immensely powerful in the future, and certainly a lot more so than people.

Power Sources

As you are likely aware, the main power source for robots right now are batteries; lead-acid batteries to be more specific. There are a lot of different battery types that can power robots. Some key factors that robotic engineers need to consider, when thinking about power supplies, is the cycle lifetime, safety, and weight of the batteries that are available. When possible, a generator can be used, like a traditional combustion engine.

Tethering a robot to an external power source takes away the need for a battery. This allows a robot to be smaller and lighter, but does not give the robot complete freedom of movement, because it must remain attached to the power source by a cable.

Here are some more power sources that could possibly be used in robotics:

- Solar power
- Nuclear
- Pneumatics
- Flywheels for storing energy
- Liquid hydraulics
- Organic waste, including garbage or even feces

Manipulators

Robots must be able to physically manipulate things, as with picking up, altering, or even destroying them. Manipulators are made up of as a robot's "hands" (otherwise known as the "end effectors") and the arm or arms (otherwise known as the "manipulators").

The effectors (hands) of a robot can usually be replaced. This allows them to be quickly repurposed for various tasks, without the need for a totally new piece of machinery. You can liken this to a person putting down a screwdriver, so they can pick up a hammer instead.

There are robots with fixed manipulators (arms), which can't be changed. Some also have a special type of manipulator that can be used with a variety of different effectors.

- **General purpose.** This type of effector is the most like the hand of a human. They are used with the most advanced types of robots, which are often humanoids. General purpose hands have a high level of dexterity, and contain hundreds of individual sensors for tactile feedback to the robot.

- **Grippers.** Mechanical grippers are one of the most widely used effector types in the robotics industry. The simplest type of gripper is made up of two "fingers", which can be opened and closed. By doing this, a robot can pick things up or release them. Some mechanical grippers are designed to work like a human hand, while others are a lot more machine like in appearance.

- **Vacuum.** These are fairly simply, but capable of holding onto big loads. However, a vacuum gripper can only manipulate an object that has a big enough, smooth surface, so that it can maintain proper suction.

Robots that are used to pick up and put down things like windows, are often designed with vacuum grippers.

Locomotion

While robots need to be able to move their own parts around, in order to carry out tasks, many of them also need to be able to move themselves entirely.

Wheels

Just to keep things simple, a lot of robots rely on a set of four wheels, as with cars and other common vehicles. Researchers do continue to work on new ways for robots to move on less wheels. This would require less manufactured parts, and give a robot a smaller footprint, so they can move into small spaces.

Robots that move on two or one wheels are known as "balancing robots". In order to stay upright without three or more wheels to keep them balanced, gyroscopes are used. These allow a robot to sense when it is beginning to go off balance, so that the wheels can be moved in order to correct that. The Segway, a personal transportation machine on two wheels, is like one of these two-wheel balancing components.

Of course, where space and using fewer parts is not a priority, some robots are made using greater than four wheels. For example, a robot with six wheels will have better grip when used in an outdoor environment.

Tracks

Tracks like those used on tanks, can give much more traction than even six, or more, wheels. That's because a set of wheels with tracks on them can operate much like hundreds of wheels would. They are used a lot in the creation of military robots, or those that are intended for outdoor use. A robot with tracks can go over some very rough and unstable terrain, without getting stuck or being put off balance. On the other hand, a tracked robot will not be able to move along smooth floors or other indoor surfaces, like carpet, very well.

Spheres

Roboticists continue to work on designs for working spherical orb robots. These are contained totally inside a ball, which is somehow moved to give the robot locomotion. This is done with a weight that spins around inside the sphere, or by actually moving the outside of the sphere itself. You will see these referred to as "ball bots" or "orb bots".

Walking

Walking is one of the more difficult issues in robotics. At this point in time, roboticists all over the world are attempting to create robots that can walk properly. There are actually a number of robots that can walk in similar ways to humans, on two legs. However, they are not nearly as capable in their movements, and they still don't walk quite like a person would.

There are robots that walk on more than just two legs. This is a much easier task to accomplish, because more legs give more balance and traction. You might wonder why people would want to use legs instead of wheels. A walking robot can go over completely uneven ground; which wheels could not possibly go over. They also offer a potentially more efficient use of energy.

As of yet, two legged robots can only walk on flat, indoor, surfaces, and even upstairs. They are not able to go over anything uneven yet.

Without going into the technical details of all of them, here are some of the different types of locomotion used in walking robots:

- Hopping
- Zero Moment Point (how the ASIMO robot moves)
- Passive dynamics
- Dynamic balancing

Climbers

There have been a lot of different methods for making robots that can climb. Human like movement has been tried, to allow a robot to move vertically up a surface that has things to grab onto. The method that geckoes use to go up walls, with the use of their toe pads, has also been copied and used in robots. This would allow a robot to go up a smooth surface, like the side of a building, for example. Another method has been to design robots that can go up anything that resembles a pole, as a snake would.

Snakes

A number of this type of robot has been created with success. By using the same types of movements that a real snake uses, a snaking robot can move through very small spaces. They might be used in the future for search and rescue missions, where people are caught inside wrecked buildings, for example. There is a Japanese snake robot that can actually move over the land or water, which is still being developed for commercial use.

Flyers

If you want to get technical about it, the type of airplane you would board to go on vacation, is a flying robot. While a pilot will be onboard for every flight, the autopilot is capable of taking off, carrying out the actual flight, and landing when a destination has been reached.

As we learned in the chapter titled The Uses of Robotics, there are also unmanned aerial vehicles (otherwise known as UAVs). Without the need for a human pilot, UAVs can be much lighter. They can also do things that would be very dangerous for a human, such as going into enemy territory in the military. Some military UAVs can automatically fire weapons at their targets, without waiting for any sort of control signal from a person. In fact, cruise missiles are basically flying robots.

Sailors

Yes, there are sailing robots that are already in use. Vaimos is a robot sailboat, which does look like an actual boat, without any people. It is moved around the ocean by wind, so it only needs battery power for the actuators in the sail and rudder, the computer, and for staying in touch with the humans at its base. If a sailboat were given reliable solar power, it could keep on going forever, in theory.

Swimmers

Scientists have worked out that some fish can achieve more than 90% efficiency with their movement through the water. They are also much better at maneuvering and picking up speed than any water craft people have ever made. They also shift the water around them less, while making less sound. It's understandable that roboticists who work on underwater robots, would like to mimic how these fish move through the water.

Skaters

There are a few robots that have been created with the ability to skate. One example is a robot with four legs that also has wheels without power. While it can walk as a normal walker, it can use these wheels to roll along in a skating motion.

Research in Robotics

While robotics has advanced leaps and bounds in recent years, there is still a lot that needs to be developed. It seems that humanity is on the cusp of uncovering those next big steps, which will hopefully produce some truly independent robots; they will be able to think for themselves, and do things without being monitored or repurposed in between tasks. Imagine how amazing the world could be, if these yet unmade technologies are used well.

While the following issues are the biggest hurdles that stand in the way between roboticists, and truly advanced robots, some other problems include:

Varying Tasks

A single robot design still can't be used to perform a range of various tasks. They can be built for a specific and they will do their job very well. In addition, they can be repurposed, as with changing out the tools that they are using, or altering their programming code. But multipurpose robots are still in their infantile state, and used more for research and development than actual, practical use. That's why you don't see robots like ASIMO doing any labor, or helping in any seriously commercial way.

Robots still cannot move around areas that have a lot of clutter, or objects that they don't recognize. In real life scenarios, they are still far behind what humans can do. They have trouble understanding speech, and certainly are not able to properly think for themselves yet. However, this gives roboticists a lot of very exciting areas in which they can research.

Making Choices

Robots are not good at making choices, unless they already have the options programmed into them. For example, imagine a robot came to a corridor with three different directions, and needed to choose one. However, it was in a building that it had never been to before, and it contained no knowledge about the layout of the area. How do you think a robot would make this type of decision? First, it helps to think about how a human might choose.

It might be possible to go back and find out which way to go, by asking someone or looking for a map. You might decide to take your chances and just choose a direction at random. Since you know the size and shape of the property, and where you came in, it might be possible to use an educated guess, or estimation, to work out which way to try. It might be that you remember something about the correct way, which you overheard earlier. These are all things that a human would consider without much hesitation. For a robot, on the other hand—they are extremely difficult to program.

Seeing

Yes, robots can be mounted with a whole range of different visual sensors. Putting something like a camera on a robot will give it the ability to "see" what's happening around it. But actually getting a robot to understand the images that are being captured with that camera, is a whole other thing.

Humans actually have very good sight, in the way they can look at something, and know what is happening. In just a tiny amount of time, you can probably look at a scene that's going on around you, and figure out just what's going on. When trying to get a robot to actually "see" what's going on around it, roboticists have come across a lot of problems.

Touch

Creating something that works like skin. Since people are covered with skin, they are able to feel what is happening around them via touch. This is incredibly precise, even though most people don't pay a whole lot of attention to it throughout the day. Just attempt to use a controller or keyboard with gloves on, and you'll see how much touch really does do to help you. Robots don't have this luxury, so they need to be designed with sensors that can give an artificial sense of "touch" to them.

In addition to being able to feel through skin, people have a lot of delicate control over their muscles. That's how a human can carry out delicate tasks. Even the clumsiest human will probably have better on-the-fly muscle control than a robot. Yes, automated robots are great at doing delicate jobs, as with electronics manufacturing robots. But these machines can often only carry out one particular task. A person can switch from one thing to the next with ease.

Language

This is naturally a big issue, since robots will need to be able to interact with humans eventually, before they can truly integrate with us. Getting a robot to understand how to use language is a big roadblock to roboticists at the moment. Knowing how to use language is a big part of solving problems, even though many people take it for granted.

Future of Robotics

What Might Become?

Do you think that robots will one day be better at everything that humans are? A lot of futurists (people who study the future) believe that it will eventually happen, with the way that things are going. Others believe that true artificial intelligence might never be possible, and that would leave robots lacking in some important areas. Without the ability to really "think" for themselves, robots will never be quite on par with humans; that is the current opinion of a lot of experts in the field anyway.

As far as being stronger, faster, or just better at performing specific tasks, robots have already got people beat in a lot of fields. There are already robots in people's houses, doing their chores for them. They are also great at taking in data and processing it, making them perfect for things like telephone call services and operating systems.

There is already a robotic soccer league, called the International RoboCup Federation. It has given itself the challenge of creating robotic players who can beat the best humans in the world, by 2050. Whether or not that actually becomes a reality, only time can tell. It takes a lot more to successfully play a sport than you might think. There's plenty of information being thrown at the players, and they need to make split second decisions the whole time. On top of that, the muscle memory and acquired physical skills that go into a competitive sport, are difficult to match for a machine. This largely comes back to being able to think for themselves.

Ethical Dilemmas

Are some of the things discussed in this book causing you to worry? There is no shortage of science fiction about robots taking over the world, trying to destroy the human race, or doing a number of other nefarious things. And then there are the ideas of how robots should be treated, and what might happen when they one day become sentient. Should a robot who can think like a person be given the same rights as a person? There's a lot of room for discussion here, and it's impossible to come up with any concrete answers. Why? For starters, this is all largely hypothetical (or theoretical at best), because there aren't actually robots advanced enough to test any theories.

Ethical Robots

As robots are being created with more and more freedom, it's becoming increasingly important to think about their ethics. Take the self driving car for example. There is a great need to control exactly what they should, and should not, do. How would you feel walking down the street, knowing that there are robots controlling vehicles right beside you? Can that programmed machine be trusted to make the "right" choice when it comes to avoiding certain obstacles?

What if a child were to suddenly run into the road, but the only way for the car to avoid the child was to swerve and hit an older couple? A person could consider what they thought was right, and act upon that very quickly. These are the types of questions that are becoming very tricky, because they're going to need real answers in the coming years.

Being intelligent does not automatically ensure that a robot is "good". There are a lot of people who are not that good, and plenty who are just horrible. If we cannot understand what makes a human go bad, should we really be trying to create robots that can think like humans? There is no certainty that a robot who is let loose to work in the world, will be able to follow the moral laws that govern what a good person does.

Roboethics

While machine ethics relates to how robots should act, roboethics is all about how people should build and use robots. While robots are basically advanced machines with complex programming, which tells them how to act, their ethical treatment isn't much of a problem. They don't think or feel, so it has not been a big deal yet. But what about robots of the future, who might be able to think and feel?

Some roboticists do not concern themselves with the ethics of creating robots. It is their job to continue to advance the technology, as they are in technical fields of research and work. There are those who do want to make sure that what they're creating is overall for the "good". Others work to create robots that will be in line with their own ethics, on a longer term basis. They wonder how their creations might affect the planet in the long run.

At this point in time, artificial intelligence (in the real sense of the word) does not exist. Yes, robots can do things that could be considered intelligence. But they're still lacking that special something that makes a living, thinking, creature unique. A human has intelligence, and at this point only another human can really understand what that means. Actually expressing and agreeing about the definition of intelligence, is a whole different topic.

Many people, in a range of different fields and industries, are concerned about the ethics involved with creation artificial intelligence. If a machine can think and even feel, should it be used like any other machine? Would that be considered slavery? How about the ethics involved in using true AI (artificial intelligence) in a military setting, where an AI is being forced to cause death and destruction? Would this be ethic, or even legal, to do to a human?

These are questions that have been covered not only by authors and movie makers throughout modern history, but many intelligent people are starting to ask the questions for real. With the looming possibility that robots are going to start to become very much human in the near future—a good roboticist might be wise to start thinking about these questions in more than a theoretical way.

Stealing Jobs?

Economists have been predicting the gradual replacement of human, by robots, in the workforce. Employment as we know it today will be completely changed, and it won't take as long as you might hope; that's what a lot of people are saying anyway.

While it's not certain how many people will end up in long term unemployment, due to robotic counterparts taking their livelihoods—the face of some industries is almost certainly going the way of the robotic workforce.

The amount of jobs in manufacturing has dropped from 25% to 10% of the total workforce in the United Station, between 1970 and now. Before the early 2000s, a drop in employment had also seen a drop in productivity. However, after that point in time, productivity began to rise while employment stayed down. There has been a growing gap between unemployment and productivity. This can only demonstrate that people's jobs are being taken by automated systems: machines. And many of those machines are robotic in nature.

> *"It is with horror, frankly, that he rejects all responsibility for the idea that metal contraptions could ever replace human beings, and that by means of wires they could awaken something like life, love, or*

rebellion. He would deem this dark prospect to be either an

overestimation of machines, or a grave offence against life."

- Karl Capek (inventor of the term "robot")

Computer Code

This chapter is going to get a little more technical, which is why it has been saved for the end of the book.

Robot Programming Languages

A programming language is just that, a language. But it is one that has been created for use by a machine; usually this is a computer. These programming languages can be used by people, to create programs (otherwise known as software). A robot needs to have a programming language that it can "understand", before any set of instructions (software) can be created for it.

Most roboticists create their own programming language for the hardware they make. This is because the machinery is so highly specialized, that premade software often isn't quite suitable for the job at hand. This does happen in other industries, but it is particularly common with robots. This does make some things difficult, and it would be easier to be able to use a standardized language, in many cases.

There are a lot of different companies making robots, and each of them has its own programming language. You might think this would be a total nightmare for computer programmers. However, there are a lot of similarities between most different programming languages. Most of the people working in this field will have a general understanding about how many of the languages work. They should be able to work with a new programming language, without actually needing to learn how to use it fluently. Roboticists will also specialize in particular programming languages, and try to work with the ones that they are best with.

Robot Software

However it is done, the purpose of making software for robots is to "show" them how to do certain things. Robotics software is the actual computer code, which tells a robot what it should do. It can be used to get a machine to perform a task automatically, as with the robots that have been discussed in this book. There are a lot of different systems that have been created, or theorized, for programming robots with more ease.

A series of programming commands is known as a piece of software. Even though this type of software has a highly specific purpose, there is still a diverse range of different types.

The common sections of a piece of software for a commercial robot include:

- Data objects
- Instructions
- Program flow (a list of instructions)

One instruction might tell the robot to move to a certain part of a work area, as with the following example:

Go to platform A

That instruction would tell the robot to move to platform A, using the rest of its coding to know what that means.

An instruction can be more complicated than this though. For example, the robot could be told to move in a more specific way:

Rotate axis 20 degrees left

This would tell the robot to turn itself to the left. You could use a program flow (list of instructions) to tell the robot how to move to a certain area of a work space. Using the software that has already been programmed for the robot, it can be told how to do new things. Of course, the robot has to be physically capable of doing so, and be using programming that's also capable enough.

Programming Errors

Safety is a big concern when programming software for a robot. There is always the possibility of errors being left in the software. Where big robots are being used in workplaces, this can prove utterly devastating. With their sheer size and power, many industrial robots could either badly injure someone, or even kill them. This is why it's always considered unsafe for a person to be in the workspace of an automated robot, even when it's turned off. It is safer to assume that a robot could suddenly move or do something unexpected, than to be sorry when an accident does occur.

Even when a particular piece of software, and the robot using it, has been thoroughly demonstrated to be error free and without malfunctions—a lot of care must be taken at all times. This is probably one reason for humans to be so hesitant about wanting to work with more robots. Would you trust your life to one, knowing how much damage it could do? This all leads back to the Ethical Dilemmas chapter of this book, and there is a lot to think about there.

Conclusion

Now that you have a more thorough understanding of the field of robotics, you might think about those machines that surround you a little differently. Will it one day be possible for robots to think just like you do? Would you want one to carry out important tasks for you, and maybe even trust your life to it? There is a lot more to building robots than just making something that can move around and do simple things. They need to be able to carry out complex tasks without error, and without human intervention.

If you are interested in getting into one of the fields related to robotics, you will certainly need to think about going to university. People spend years studying the basics, so that they can one day create their own robots. Imagine how cool it would be to see a robotic machine that you thought up and made, doing important work, or perhaps even helping to save lives. That could happen, and even more, but you have to start somewhere.

The next step would be looking into the construction of basic robots. It's important to get some real experience, so that you can learn how they work. There are plenty of robot kits out there, which are intended for use in an educational setting. You can also work with robots in a software setting, on your computer.

Helpful Links

These links will take you to some amazing websites, which are dedicated to robotics. They're great for beginners too!

- www.letsmakerobots.com
- www.the-nref.org (The National Robotics Education Foundation)

- www.robotbooks.com (a store selling robotics books, magazines, movies, toys, and robot kits)
- Lego Mindstorm (real robot kits and products made by Lego)